I0045724

Organic Chemistry

Brick by Brick

Compound 1

Dmitry Vostokov

Organic Chemistry Brick by Brick, Compound 1: Using LEGO® to Teach Structure and Reactivity

Published by OpenTask, Republic of Ireland

Copyright © 2020 by OpenTask

Copyright © 2020 by Dmitry Vostokov

Contributions by Ekaterina Vostokova and Kirill Vostokov

All rights reserved. No part of this book may be reproduced, stored in a retrieval system, or transmitted, in any form or by any means, without the prior written permission of the publisher.

Product and company names mentioned in this book may be trademarks of their owners.

OpenTask books and magazines are available through booksellers and distributors worldwide. For further information or comments, send requests to press@opentask.com.

A CIP catalog record for this book is available from the British Library.

ISBN-13: 978-1912636020 (Paperback)

Revision 1.00 (June 2020)

Preface

During the Summer of 2018, I was thinking about the possibility of teaching artificial chemistry to early secondary school pupils before official Chemistry subject years. Before, I used bricks to represent some simple data structures and software logs[1]. Out of this synthesis, in November 2018, the idea of using a baseplate representation of chemical structures was born. Initially, I made about 250 LEGO models of various organic structures, later annotated and made their pictures available for download, described the proposal in greater detail, and extended presentation possibilities to include polymers and inorganic materials[2]. After almost two years passed since the initial invention, I decided to produce a consistent book series, including chemical reactions in addition to structures. In comparison to free slides that were created to showcase the possibilities, the book is organized differently, uses several baseplate representations including 3D models, incorporates lessons learned during modeling of machine learning[3] and category theory[4] concepts, and aims to teach organic chemistry from the ground up. For this part (I call it "compound" according to chemical terminology), I used the following books as a reference:

- Basic Principles of Organic Chemistry, 2nd edition by Roberts and Caserio
- Fundamentals of Organic Chemistry, Volume 1, 2nd edition by A.N. Nesmeyanov and N.A. Nesmeyanov.

These were my favorite textbooks during my secondary school years in the 1980s. I plan to add more references to subsequent compounds as I have accumulated a substantial library in recent years.

Some basic chemistry knowledge at a secondary school level is preferable but not necessary to follow this book.

---

[1] https://www.dumpanalysis.org/lego-log-analysis

[2] https://www.opentask.com/chemistry-brick-by-brick-series

[3] https://www.dumpanalysis.org/machine-learning-brick-by-brick-series

[4] https://www.dumpanalysis.org/visual-category-theory

Molecular formulas of compounds show the kinds and the number of atoms. We use square 2x2 bricks for individual atoms.

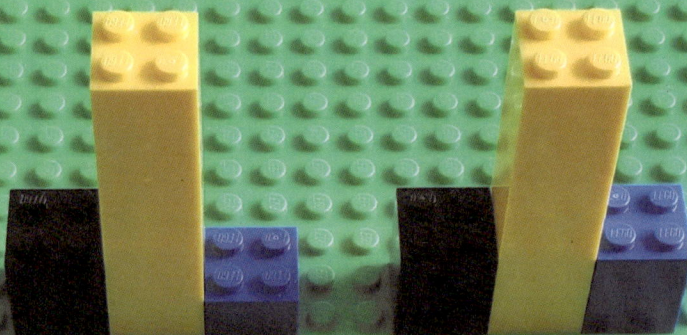

C_2H_6O

$C_2H_6O_2$

C_2H_5Br

We can hypothesize different possible structural formulas for molecular formula C_2H_5Br. We use 2x1 bricks for chemical bonds.

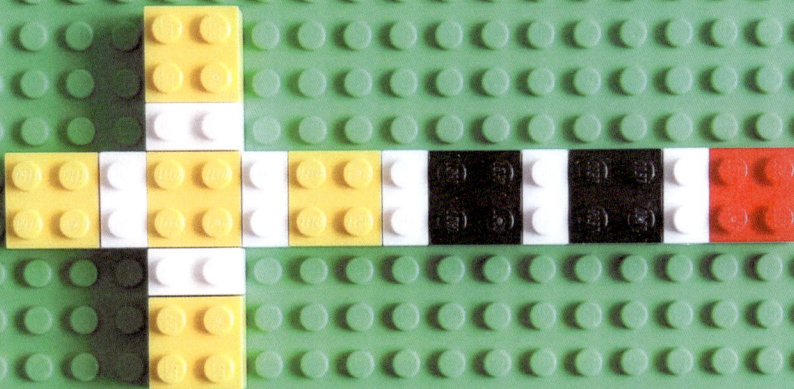

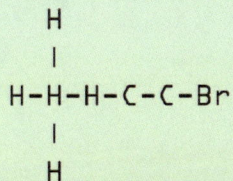

```
            H
            |
H-H-H-C-C-Br
            |
            H
```

Is this structure valid?

The following molecular formulas were determined for primitive compounds involving C (carbon), H (hydrogen), and Br (bromine) atoms.

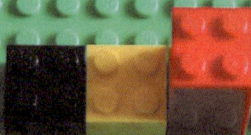

H_2

Br_2

CH_3Br

CH_2Br_2

$CHBr_3$

CBr_4

We can deduct valences (the number of bonds): C atoms are tetravalent (4 bonds), H and Br atoms are univalent (1 bond).

Based on the number of valences, we can propose the following structural formula for C_2H_5Br.

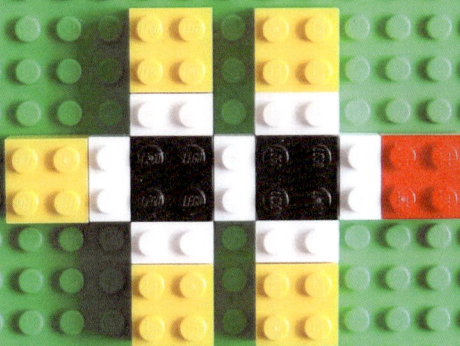

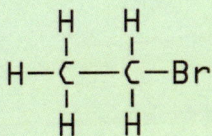

$$
\begin{array}{ccccc}
 & H & & H & \\
 & | & & | & \\
H - & C & - & C & - Br \\
 & | & & | & \\
 & H & & H &
\end{array}
$$

We can present the structural formulas for C_2H_5Br differently. These are the same structures.

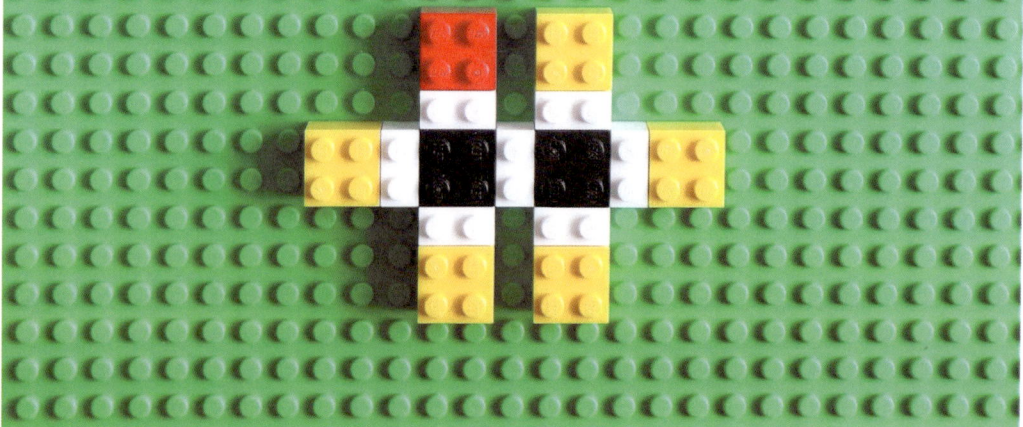

These are also different ways of showing the same structure for the C_2H_5Br molecule.

However, for these two structures, Br atom is positioned differently to the opposing CH₃ group.

Are these compounds different?

In 1874, Jacobus Henricus van 't Hoff Jr. proposed that 4 valences of carbon atom C are equivalent, and its end atoms are arranged spatially at the end of a tetrahedron.

CH₃

We see that the location of Br atom is irrelevant, and structures from the previous page are equivalent.

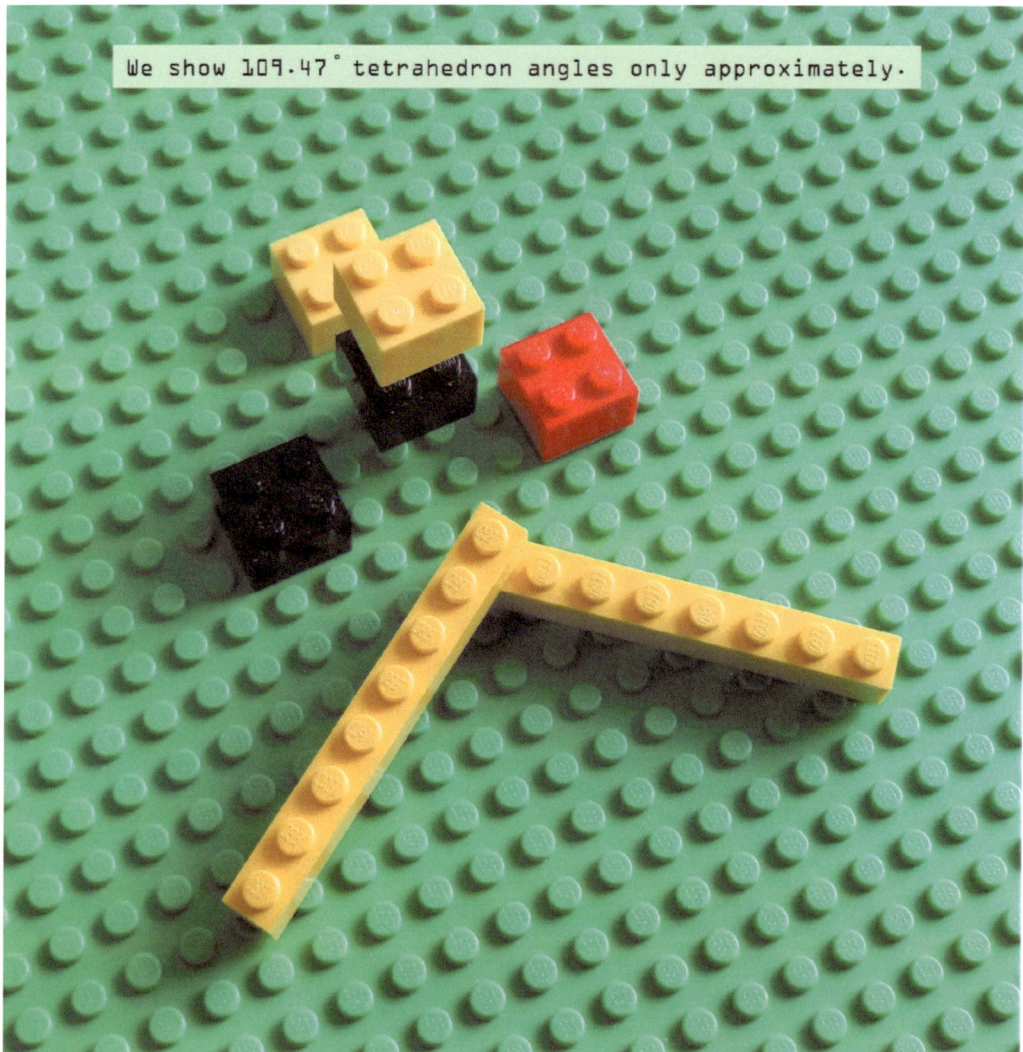

We show 109.47° tetrahedron angles only approximately.

But we can see the difference from 90°.

For the $C_2H_4Br_2$ molecule, we have a possibility of (structural) isomers: different substances with the same molecular formula.

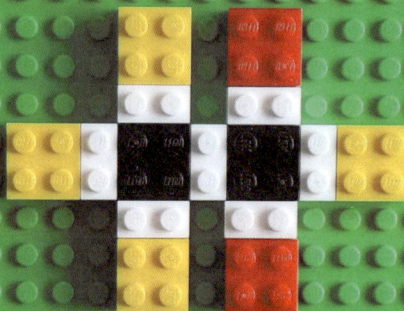

These two structures are different because the second Br atom is connected to different C atoms.

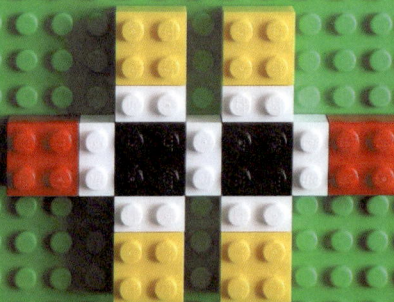

Are these structures for one $C_2H_4Br_2$ isomer equivalent?

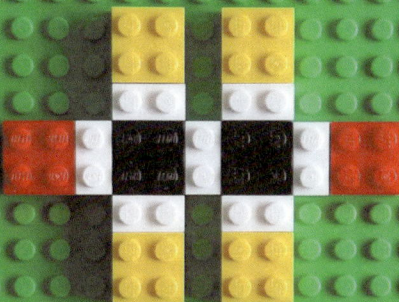

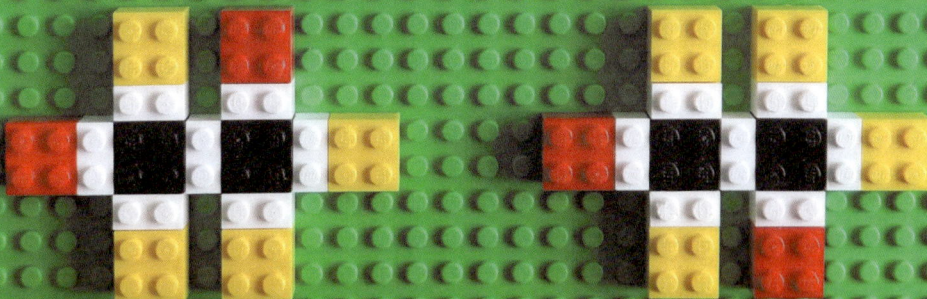

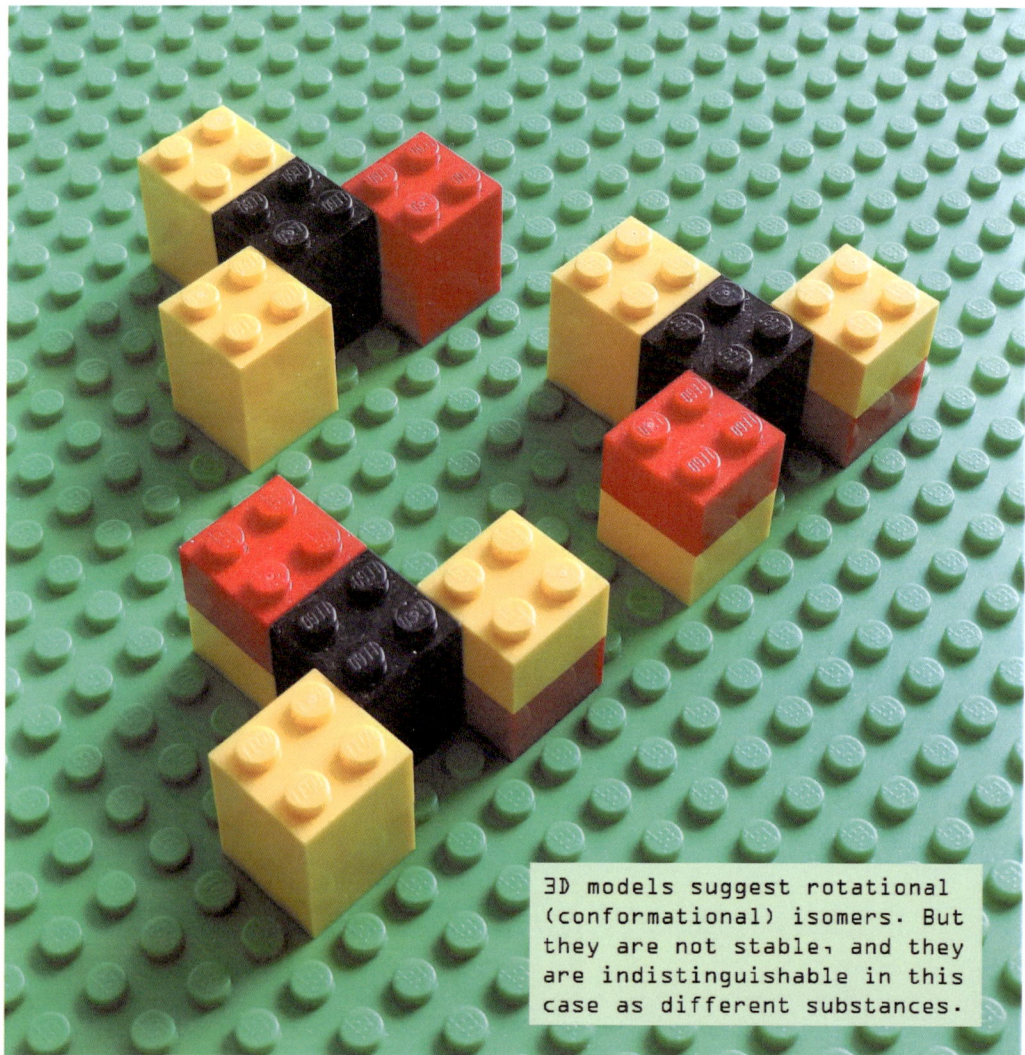

3D models suggest rotational (conformational) isomers. But they are not stable, and they are indistinguishable in this case as different substances.

Given a substance with the molecular formula, how can we determine its structural formula out of several possibilities?

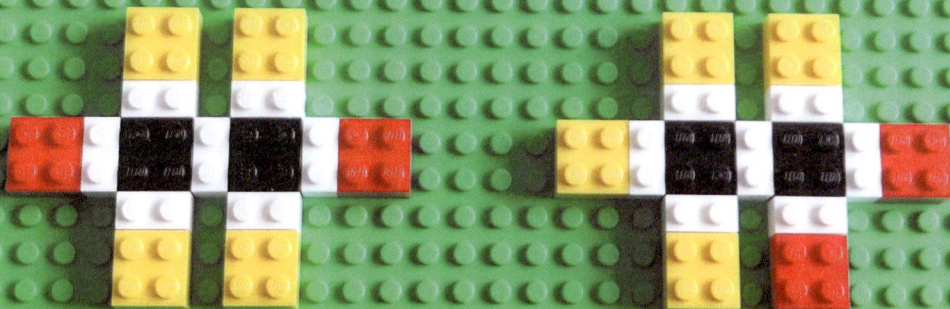

$C_2H_4Br_2$

This?

Or this?

For this structure, if we substitute a hydrogen atom by another bromine Br atom, we get only one new substance.

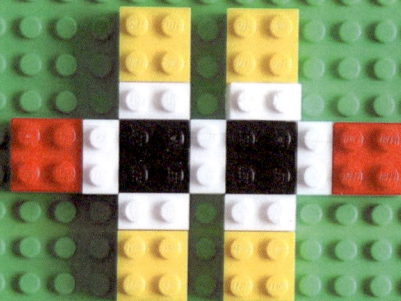

$C_2H_4Br_2$

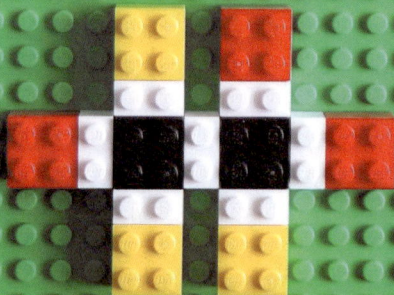

$C_2H_3Br_3$

However, for this structure, if we substitute a hydrogen atom by another bromine Br atom, we get two different substances.

$C_2H_4Br_2$

$C_2H_3Br_3$

$C_2H_3Br_3$
different
properties

Let's look at how a structure is determined through chemical reactions. C_2H_5Br reacts with water H_2O to produce C_2H_6O.

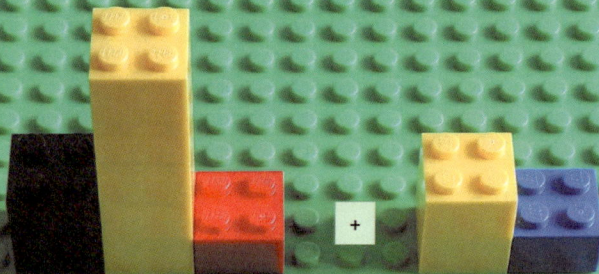

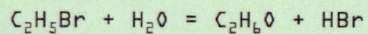

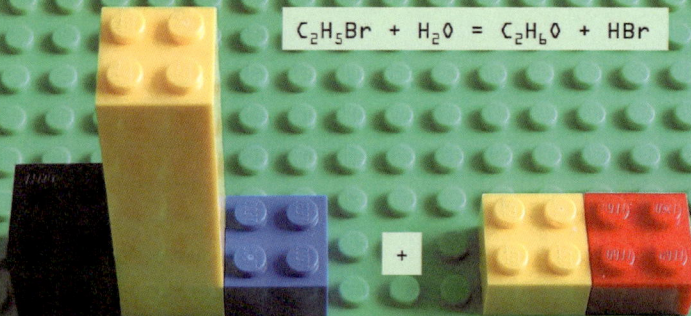

$$C_2H_5Br + H_2O = C_2H_6O + HBr$$

Regarding the reaction, it is the equality if we count atoms on the left and the right side of the equation.

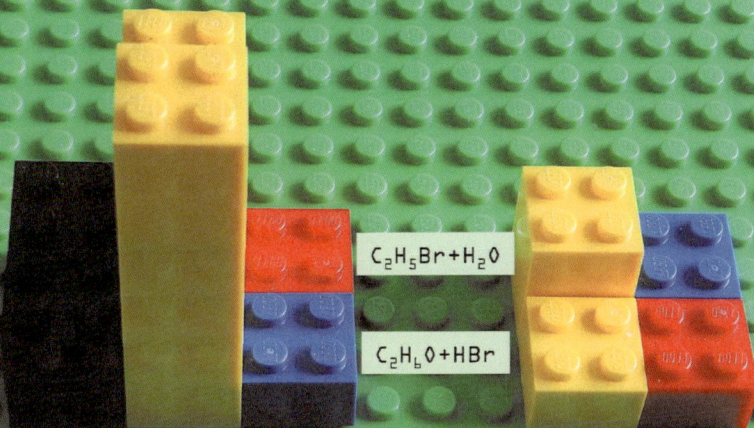

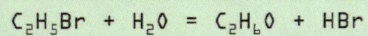

$C_2H_5Br+H_2O$

C_2H_6O+HBr

$$C_2H_5Br + H_2O = C_2H_6O + HBr$$

However, we want to find out which isomer structure is produced. The principle of least structural change suggests the left structure.

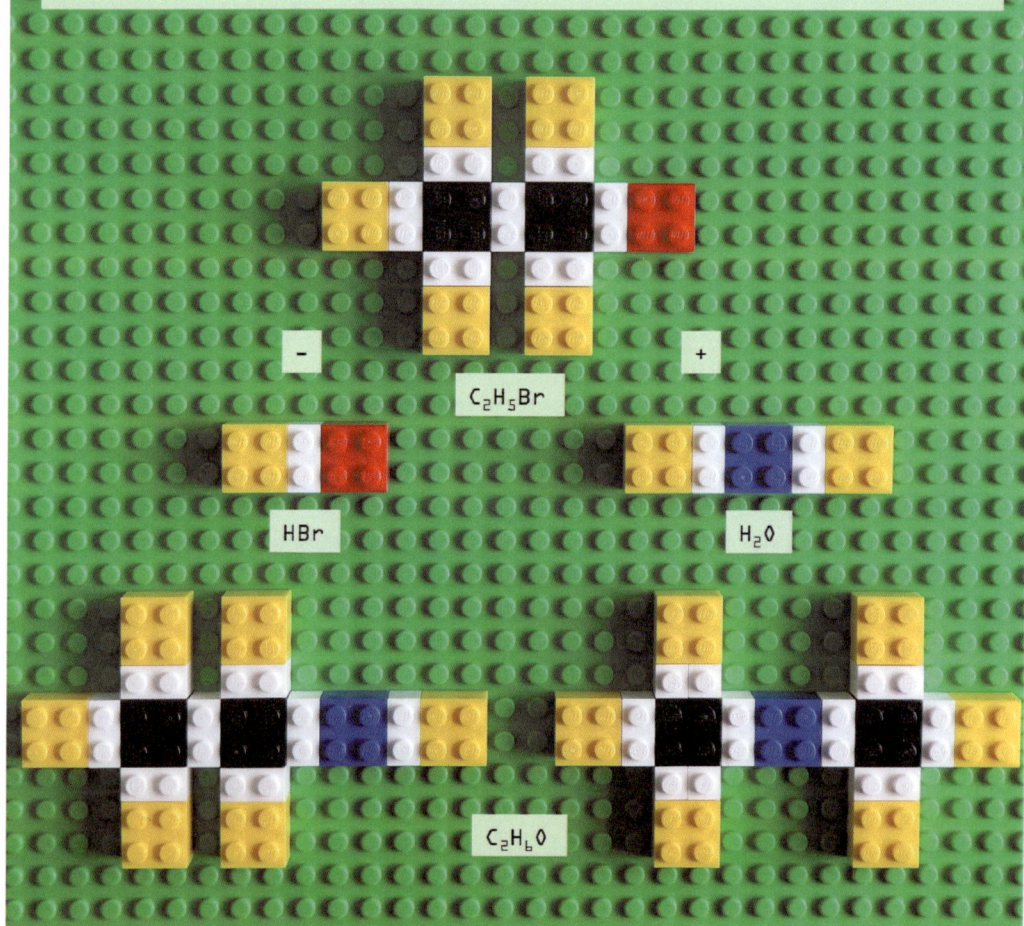

C_2H_5Br

−   +

HBr H_2O

C_2H_6O

Indeed, the left structure reacts with HBr to produce the original structure back.

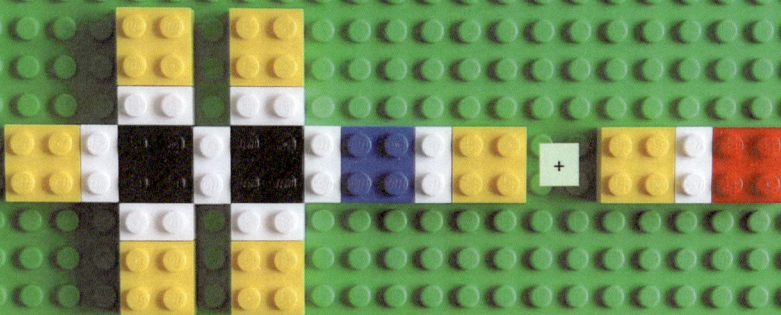

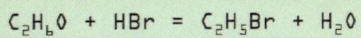

$$C_2H_6O + HBr = C_2H_5Br + H_2O$$

But the right isomer structure reacts with 2 molecules of HBr to produce 2 molecules of CH_3Br.

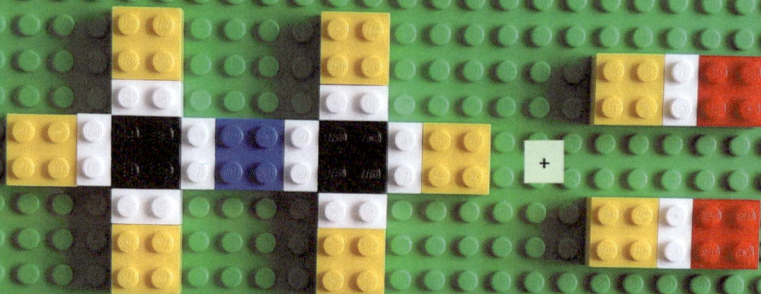

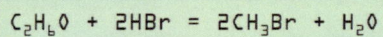

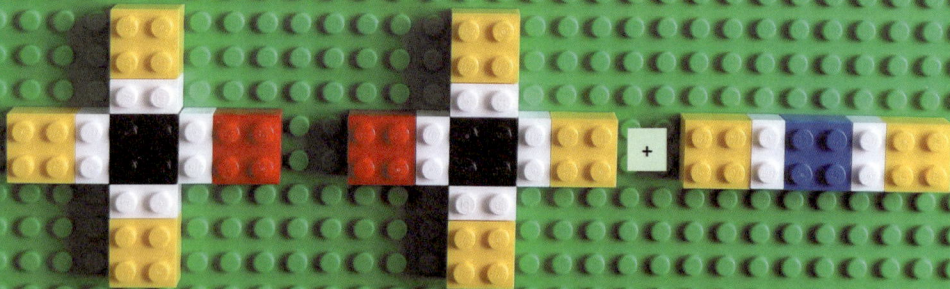

$$C_2H_6O + 2HBr = 2CH_3Br + H_2O$$

There are exceptions to the principle of least structural change. For example, the following reaction is forbidden.

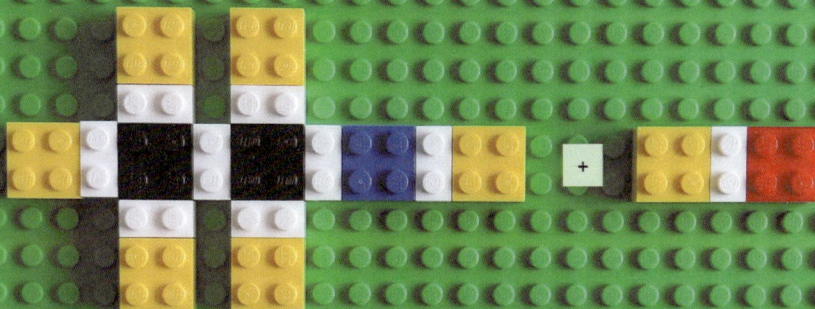

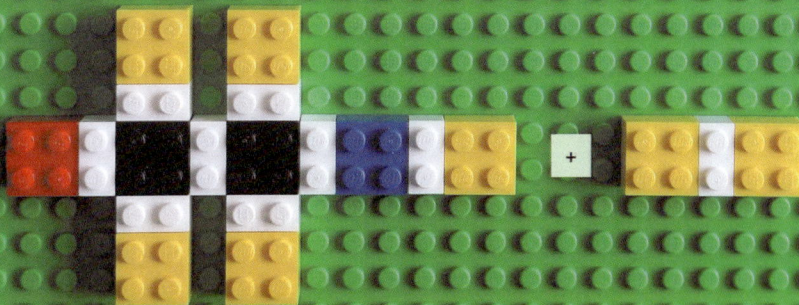

The molecular formula of benzene C_6H_6.

In 1866, August Kekulé suggested the following structure: carbon atoms C form a hexagonal ring with alternating single and double bonds (shown as orange 2x1 bricks) with each carbon atom connected to one hydrogen atom H.

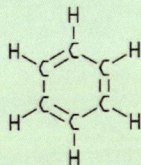

We see there is a possibility of different benzene structures
(we hide hydrogen atoms for clarity).

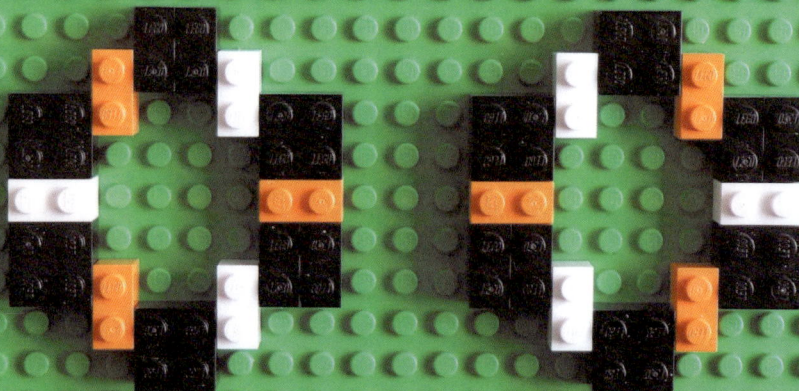

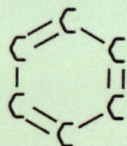

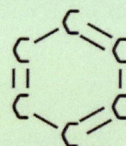

Also, the substitution of 2 adjacent hydrogen atoms by bromine Br atoms suggests 2 different structures as well (we hide the remaining hydrogen atoms for clarity).

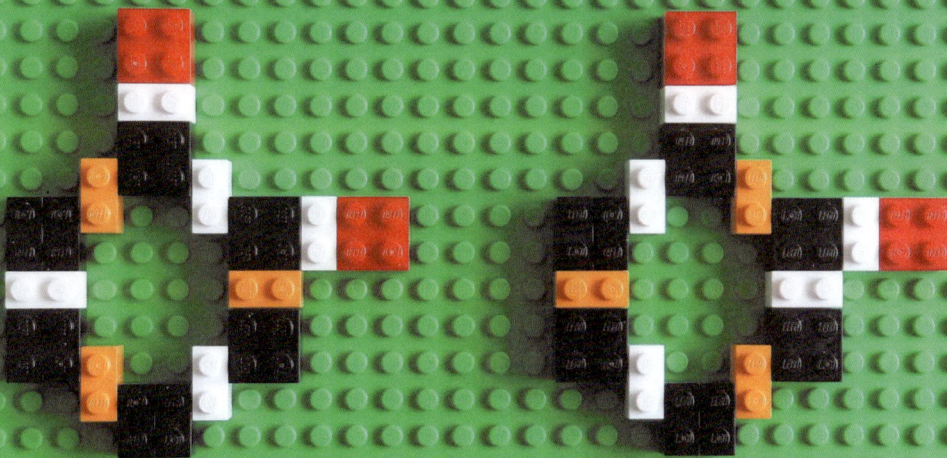

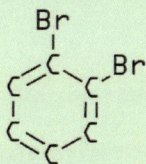

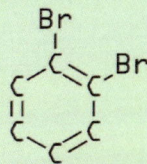

Kekulé explained this by a rapid equilibrium of 2 different structures. We model this by using a small square inside the carbon hexagon. Hydrogen atom may also be omitted.

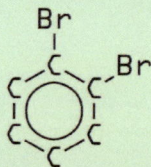

Br
C
C   Br
C   C
C   C
C

It was later found that all C-C bonds in benzene are equivalent.

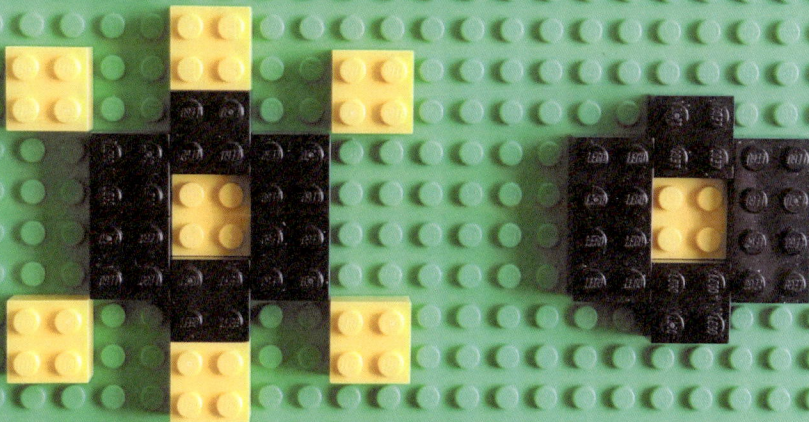

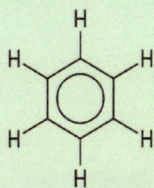

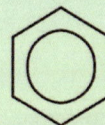

It is possible to represent structures differently. For example, in hydrogen cyanide, we can also show a triple bond differently and even omit bonds entirely. This is useful for building complex structures.

HCN

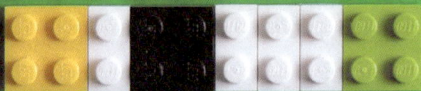

H—C≡N

Molecular formulas and structures for ethylene C_2H_4 and acetylene (ethyne) C_2H_2.

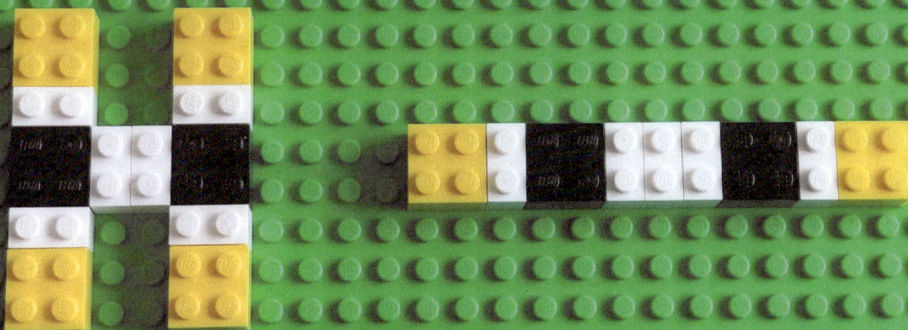

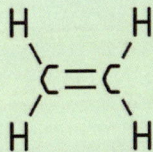

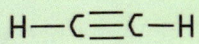

$$\begin{array}{cc} H & H \\ & \diagdown \quad \diagup \\ & C = C \\ & \diagup \quad \diagdown \\ H & H \end{array}$$

$$H - C \equiv C - H$$

For visual clarity, double and triple bonds are shown using orange and red 2x1 bricks.

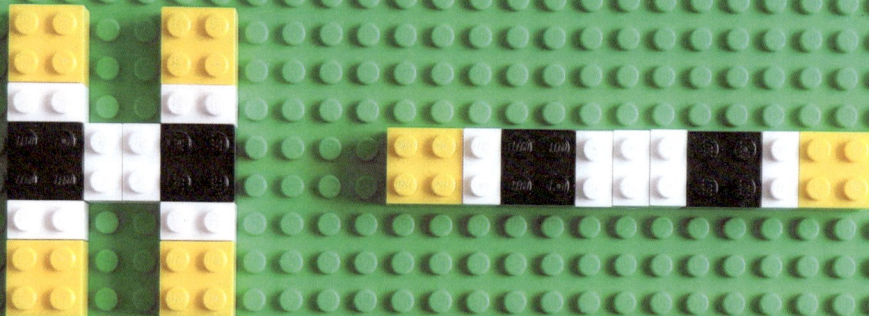

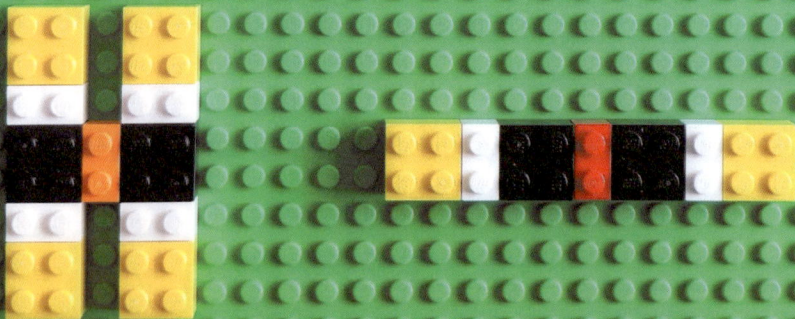

In 1828, Friedrich Wöhler synthesized organic compound urea from inorganic ammonium cyanate. It is considered as the starting point of modern organic chemistry.

CH_4N_2O

NH_4CNO → NH_2CONH_2

Here we have semi-structured molecular formulas.

Let's look at the structural formulas for ammonium cyanate NH_4CNO and urea (carbamide) NH_2CONH_2. We omit bonds here.

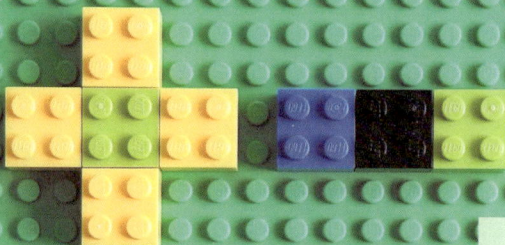

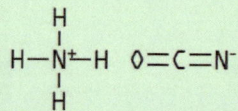

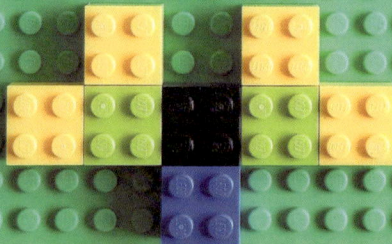

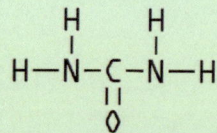

$$H-\overset{\displaystyle H}{\underset{\displaystyle H}{\overset{|}{\underset{|}{N^+}}}}-H \qquad O=C=N^-$$

$$H-\overset{\displaystyle H}{\overset{|}{N}}-\underset{\displaystyle \underset{O}{\overset{||}{}}}{C}-\overset{\displaystyle H}{\overset{|}{N}}-H$$

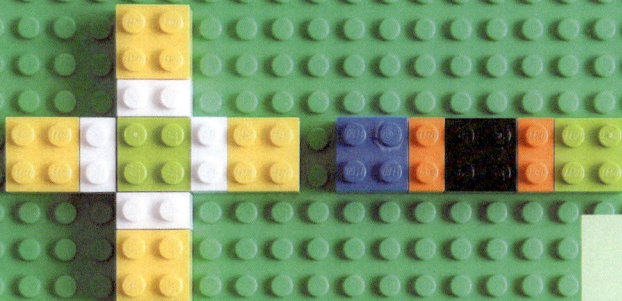

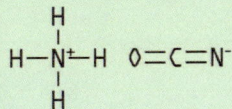

$$H-\overset{\overset{\displaystyle H}{|}}{\underset{\underset{\displaystyle H}{|}}{N^+}}-H \qquad O=C=N^-$$

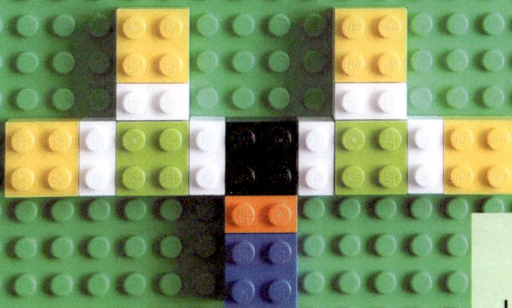

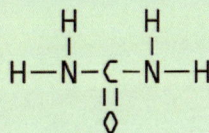

$$H-\overset{\overset{\displaystyle H}{|}}{N}-\overset{\overset{\displaystyle}{\underset{\underset{\displaystyle O}{||}}{C}}}-\overset{\overset{\displaystyle H}{|}}{N}-H$$

We may also leave single bonds out as we did
originally with baseplate representation.

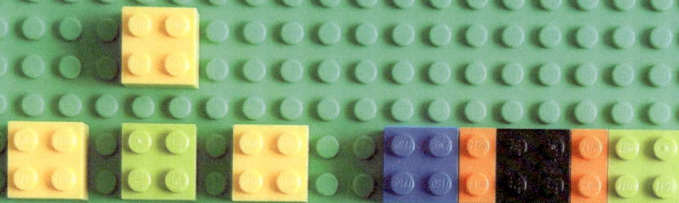

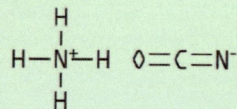

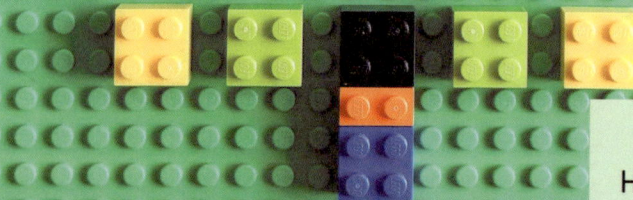

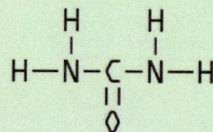

$$H-\overset{\overset{\displaystyle H}{|}}{\underset{\underset{\displaystyle H}{|}}{N^+}}-H \qquad O=C=N^-$$

$$H-\overset{\overset{\displaystyle H}{|}}{N}-\overset{\overset{\displaystyle }{\underset{\underset{\displaystyle O}{||}}{C}}}-\overset{\overset{\displaystyle H}{|}}{N}-H$$

The same structures using a more compact representation that we now prefer.

This is an original baseplate representation with implicit hydrogen atoms.

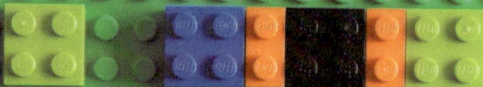

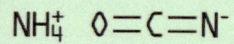

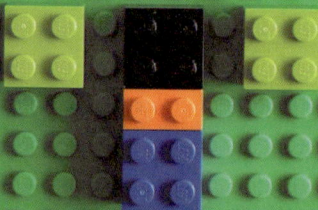

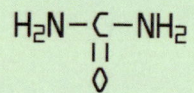

$$NH_4^+ \quad O{=}C{=}N^-$$

$$H_2N-\underset{\underset{O}{\|}}{C}-NH_2$$

This is an implicit baseplate representation that we prefer now.

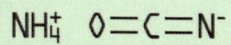

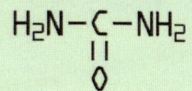

NH_4^+ $O{=}C{=}N^-$

$$H_2N-\underset{\underset{\textstyle O}{\|}}{C}-NH_2$$

41

We can rebuild models with implicit single bonds and hydrogen atoms for $C_2H_4Br_2$ and C_2H_6O isomers.

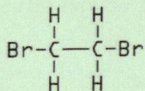

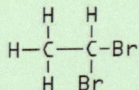

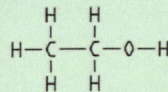

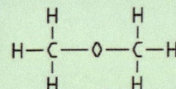

```
   H  H
   |  |
Br-C--C-Br
   |  |
   H  H
```

```
   H  H
   |  |
 H-C--C-Br
   |  |
   H  Br
```

```
   H  H
   |  |
 H-C--C-O-H
   |  |
   H  H
```

```
   H       H
   |       |
 H-C---O---C-H
   |       |
   H       H
```

These are more compact structural models for the same molecules.

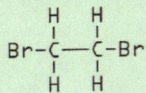

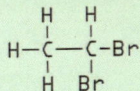

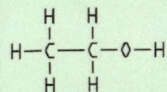

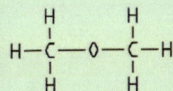

$$
\begin{array}{ccc}
& H & H \\
& | & | \\
Br-&C-&C-Br \\
& | & | \\
& H & H
\end{array}
$$

$$
\begin{array}{ccc}
& H & H \\
& | & | \\
H-&C-&C-Br \\
& | & | \\
& H & Br
\end{array}
$$

$$
\begin{array}{ccc}
& H & H \\
& | & | \\
H-&C-&C-O-H \\
& | & | \\
& H & H
\end{array}
$$

$$
\begin{array}{ccc}
& H & & H \\
& | & & | \\
H-&C-&O-&C-H \\
& | & & | \\
& H & & H
\end{array}
$$

43

www.ingramcontent.com/pod-product-compliance
Lightning Source LLC
Chambersburg PA
CBRC101955190326
41519CB00006B/279